DR. JACK

The Quantum Weaver

Untangling the Threads of Reality

First edition

This book was professionally typeset on Reedsy.
Find out more at reedsy.com

Contents

Preface

I have always been jealous of magicians.

A magician walks onto a stage, rolls up his sleeves, and performs the impossible. He saws a person in half. He makes an elephant disappear. He pulls a rabbit from an empty hat. We watch, delighted, because we know it is a trick. We know that behind the smoke and mirrors, there is a trapdoor or a hidden wire. We know the world is actually solid and sensible, and the magic is just a temporary illusion.

But what if the magician wasn't tricking you?

What if the woman really *was* in two halves? What if the elephant *did* vanish into thin air and reappear somewhere else instantly? What if the rabbit didn't exist until the magician reached into the hat to grab it?

This is the reality of Quantum Mechanics.

For the last century, physicists have been peering under the curtain of our reality, expecting to find the trapdoors and wires—the "gears" of the universe that explain how it all works. Instead, they found something that defies all reason. They found a world where objects can be in two places at once, where cause and effect can be reversed, and where the act of looking at something changes what it is.

I wrote this book because I believe this "magic" shouldn't belong only to the mathematicians scribbling Greek letters on chalkboards. The story of the quantum world is the story of us. It is the story of what we are made of. It is the most dramatic, confusing, and beautiful narrative in the history of human thought.

You do not need a degree in physics to read this book. You do not need to know calculus. You just need to be willing to let go of your common sense for a few hours.

We are going to pull back the curtain. I warn you now: there are no wires.

Introduction

The Loom of Existence

Look at your hand.

Hold it up in front of your face. Flex your fingers. Make a fist.

It looks solid, doesn't it? It feels continuous. If you slam it against the table, it makes a satisfying *thud.* It stops. It doesn't pass through the wood like a ghost. To your eyes and your touch, your hand is a dense, unbroken landscape of skin, bone, and blood.

But this solidity is a magnificent lie.

If you were to shrink yourself down—smaller than a cell, smaller than a DNA helix, down to the scale of an atom—you would find that your hand is almost entirely empty space.

To be precise, an atom is **99.9999999999996%** void. To visualize this, imagine the nucleus of an atom (where almost all the mass is) as a marble sitting on the center line of a football stadium. The electrons—the things that define the "size" of the atom—would be tiny gnats buzzing in the highest seats of the stands.

Everything in between the marble and the gnats is empty.

This means that if you removed all the empty space from the atoms that make up all the humans on Earth, you could fit the entire population of the planet inside a single sugar cube.

So, here is the question that should keep you up at night: **Why don't you fall through the floor?**

If you are mostly empty space, and the floor is mostly empty space, why do you stand on it? Why doesn't your hand pass through the table like smoke through a screen door?

The answer lies in the **Weaver**.

There are invisible fields electromagnetic forces and exclusion principles that knit this empty space together. We are not made of "stuff" in the traditional sense; we are made of vibrations. We are patterns in a cosmic fabric, woven tight by rules that defy our common sense.

The Broken Clock

For a long time, we thought we knew the rules. In the classical world of Isaac Newton, the universe was a clockwork machine. It was predictable. If you kicked a ball, you could calculate exactly where it would land. If you heated water, you knew exactly when it would boil. The math was deterministic.

$F = ma$

Force equals mass times acceleration. Simple. Elegant. Safe.

But in the early 20th century, physicists pulled a loose thread on this tapestry, and the whole picture unraveled. They discovered that at the bottom of reality down in the basement where the atoms live the rules change.

Certainty is replaced by probability.

Particles can be in two places at once.
Time can flow backward.
Objects can communicate instantly across light-years.
Opens in a new window Shutter stock

This layer of reality is the **Quantum Realm**. It is the loom upon which our universe is woven.

It is a world where a cat can be both dead and alive until you check on it. It is a world where a particle can walk through a solid wall if it hits it enough times. It is a world where the mere act of watching an experiment changes the outcome.

In this book, we will descend into that microscopic world. We will meet the weavers—the particles and forces and we will try to understand the strange, shimmering tapestry they create. We will look at the equations that describe this chaos, like the famous uncertainty principle:

Delta x Delta $p \geq h/2$

This inequality tells us that nature has a fuzziness limit. You can know where a particle is, or how fast it is going, but never both perfectly at the same time. The Weaver refuses to be pinned down.

Welcome to the fabric of reality. Watch your step. The floor is not as solid as it looks.

The Ultraviolet Catastrophe

In the year 1874, a polite, bespectacled teenager named Max Planck walked into the office of his professor, Philip von Jolly, at the University of Munich. Max was a talented musician—he played the piano, the cello, and had perfect pitch—but he was drawn to the quiet, absolute truths of mathematics. He wanted to know if he should dedicate his life to theoretical physics.

Professor von Jolly looked at the boy with a kind of pitying exhaustion and gave him advice that has since become the most famous bad take in the history of science.

"Physics," the professor said, *"is essentially a finished system. All the fundamental laws have been discovered. There is nothing left to do but fill in a few decimal places and catalog the minor details."*

It is easy to laugh at von Jolly now, but at the time, he was right. If you were a scientist in late Victorian Europe, the world made perfect sense. The universe was a Masterpiece of Order. Isaac Newton had given us the laws of motion and gravity, explaining how everything from falling apples to orbiting planets behaved. James Clerk Maxwell had recently unified electricity and magnetism into a set of beautiful, flowing equations. Thermodynamics explained engines and heat.

We had the blueprints of reality. The universe was a Great Machine a clockwork mechanism of gears and levers that ticked along deterministically. If you knew the position and speed of every particle in the universe, you could, in theory, predict the future forever. There were no mysteries left. Just cleanup work.

But there were two small problems. Two tiny, irritating "clouds" on the horizon, as Lord Kelvin called them, that refused to fit into this perfect picture. One of those clouds was the speed of light. The other was the color of toast.

And that second cloud was about to unleash a storm that would tear the Great Machine apart.

The Color of Heat

The problem began in the factories of the German Empire. It was the 1890s, the height of the industrial revolution, and Germany was obsessed with efficiency. The lighting industry was booming, and companies like Siemens wanted to know how to build the perfect light bulb and the most efficient blast furnace.

To do this, they needed to understand the relationship between **heat** and **light**.

We all understand this relationship intuitively. If you put an iron poker into a blacksmith's fire, nothing happens at first. Then, as it gets hotter, it glows a dull, angry red. Make it hotter, and it turns bright orange. Hotter still, and it becomes a blinding yellow-white. If you could heat it further without melting it, it would eventually glow blue.

This is the universal code of matter: **As temperature rises, the frequency of the light it emits rises.** Red is a low-frequency wave; blue is a high-frequency wave.

Physicists called a perfect absorbed and emitter of this energy a "Black Body." They wanted to write a mathematical formula that could predict exactly how much light of each color a black body would emit at a given temperature. It seemed like a simple task. They had all the laws of thermodynamics and electromagnetism. They just had to do the math.

But when the best minds in Europe sat down to calculate it, the equation broke.

The Catastrophe

Two British physicists, Lord Rayleigh and James Jeans, took a crack at it using the standard rules of classical physics. They assumed quite logically that energy is a continuous thing, like water flowing from a tap. They assumed that a hot object could vibrate at any frequency it wanted.

Their equation worked beautifully for low-frequency light (infrared and red). But as they moved up the spectrum toward the higher frequencies blue, violet, and ultraviolet the math went insane.

According to the "Rayleigh-Jeans Law," as the frequency of light increased, the energy emitted should increase exponentially. The curve on the graph didn't come back down; it shot straight up.

This wasn't just a minor error. It was a physical impossibility.

The equation predicted that if you opened your oven door to check on a roast chicken, the oven shouldn't just bathe you in warm infrared light. It should blast you with ultraviolet radiation, then X-rays, then Gamma rays. It should release an infinite amount of energy in a split second, vaporizing you, the kitchen, and the entire city.

According to the best physics of the 19th century, every time you lit a match, the universe should end.

They called this **The Ultraviolet Catastrophe**.

Obviously, the universe was not ending. Ovens were safe. Light bulbs worked. But the fact that they worked meant that the laws of physics were wrong. The Great Machine was broken.

The Act of Desperation

Max Planck was now a grown man, a professor at the University of Berlin. He was not a revolutionary. He was a Prussian traditionalist who loved the old order. He revered Newton and Maxwell. The chaos of the Ultraviolet Catastrophe bothered him deeply, not because he wanted to discover something new, but because he wanted to restore order to the old system.

For years, Planck wrestled with the data. He had the experimental results from the bureau of standards he knew what the curve *should* look like (it should go up, peak, and come back down). But he couldn't make the theory match the reality.

In late 1900, Planck did something that went against every instinct he had. He decided to "cheat."

He looked at the math and asked a strange question: *What if energy isn't like water? What if it's like sand?*

In the classical world, energy was viewed as continuous. You could have 1.5 Joules, or 1.55555 Joules, or any amount you pleased. But Planck wondered what would happen if he forced the math to treat energy as discrete packets—little "chunks" of energy that couldn't be divided.

He realized that high-frequency light (like ultraviolet) is "expensive." It takes a lot of energy to make a wave that vibrates that fast. Low-frequency light (like red) is "cheap."

If energy comes in packets, then at high temperatures, it's easy for the oven to spit out millions of "cheap" red packets. But it's incredibly difficult for the oven to save up enough energy to buy even a single "expensive" ultraviolet packet.

Because the oven can't "afford" the high-frequency light, the curve drops back down. The catastrophe is avoided. The infinite energy blast is stifled by the high cost of the admission ticket.

Planck wrote down a simple formula to describe the size of these packets:

$$E = hf$$

Energy (E) equals the frequency (f) times a tiny, fundamental number that would later be known as **Planck's Constant (*h*)**.

The Reluctant Father

Planck introduced his idea on December 14, 1900. The math was perfect. It matched the experimental data exactly. The Ultraviolet Catastrophe vanished, replaced by a smooth, elegant curve.

But here is the twist: **Max Planck didn't believe it.**

He didn't think he had discovered a fundamental truth about reality. He thought he had found a "mathematical trick" a convenient fudge factor that made the equations work until a real physicist could find the proper explanation.

He spent the next several years trying to get rid of his own discovery. He tried to smooth the "sand" back into "water." He desperately wanted the world to be continuous again. He couldn't accept the implication of his own equation: that the universe is pix elated.

This is the great irony of the quantum revolution. It wasn't started by a rebel trying to overthrow the government. It was started by a conservative bureaucrat who accidentally blew up the building while trying to fix the plumbing.

Planck had opened the door to the quantum realm, but he refused to walk through it. He stood on the threshold, apologizing for the mess. It would take someone much younger, much brasher, and much less respectful of authority to push the door wide open.

The Schizophrenic Photon

While Max Planck was wringing his hands in Berlin, worrying about the damage he had done to classical physics, a twenty-six-year-old clerk was sitting in a patent office in Bern, Switzerland, preparing to destroy it completely.

The year was 1905. The clerk was Albert Einstein.

He was a nobody. He had failed to get a job in academia. His doctoral thesis had been rejected. He spent his days sitting on a high stool, examining blueprints for gravel sorters and electromagnetically devices. But while his boss wasn't looking, Einstein was daydreaming about light.

In that single year—his *Annus Mirabilis* or "Miracle Year"—Einstein published four papers that would alter the course of human history. One proved the existence of atoms. One introduced Special Relativity ($E=mc^2$). But the one that would eventually win him the Nobel Prize was the one that took Planck's "mathematical trick" and turned it into physical reality.

The War of the Waves

To understand why Einstein's idea was so radical, we have to look at the battlefield of physics regarding light. For centuries, a war had raged: **Is light a particle or a wave?**

Isaac Newton, the giant of physics, had insisted light was made of particles—tiny "corpuscles" like microscopic bullets shooting from a candle. Because Newton was Newton, everyone believed him for a hundred years.

But in the early 1800s, a polymath named Thomas Young performed an experiment that seemed to kill Newton's idea dead. Young passed light through two narrow slits and projected it onto a screen. If light were made of bullets, you would expect two bright lines on the screen. Instead, Young saw a series of alternating bright and dark stripes—an **interference pattern**.

This is exactly what water does. If you drop two stones in a pond, the ripples overlap. Where two crests meet, the water gets higher (constructive interference). Where a crest meets a trough, they cancel out and the water goes flat (destructive interference).

Young proved that light rippled. James Clerk Maxwell later provided the math, showing that light was an electromagnetic wave. By 1900, the debate was over. Newton was wrong. Light was a wave. Case closed.

But there was a loose thread. A phenomenon called the **Photoelectric Effect**.

The Knockout Punch

Scientists had noticed that if you shine a light on a piece of metal, like zinc, the light can knock electrons off the surface. It's like spraying a dirty car with a hose; the water (light) knocks the mud (electrons) loose.

If light is a wave, the rules should be simple:

1. **Intensity (Brightness) equals Power.** A brighter light is a bigger wave (a tsunami). It should smash into the electrons harder and send them flying with more speed.
2. **Frequency (Color) shouldn't matter much.** A big wave is a big wave,

whether the ripples are tight or spaced out.

But when experimental actually tested this in the lab, the universe refused to cooperate.

- They turned up the brightness (the intensity) to blinding levels. More electrons fell off, sure, but they came off at the same lazy speed. The "tsunami" wasn't hitting any harder.
- Then, they changed the color to dim blue or ultraviolet. Suddenly, the electrons flew off like bullets.

This made no sense in the wave model. Why would a tiny, dim ripple of blue light pack more of a punch than a giant, blinding tsunami of red light?

Einstein saw the answer. He looked at Planck's equation, $E=hf$, and realized it wasn't just about how ovens emit heat. It was about the light itself.

Einstein proposed that light is **not** a continuous wave. He said light is a stream of tiny, discrete bullets. Today, we call them **photons**.

In Einstein's view, the "color" of the light tells you how heavy the bullet is.

- **Red light** is a stream of ping-pong balls. You can fire a million of them (high brightness), but they are too light to knock an electron loose. They just bounce off.
- **Blue light** is a stream of cannonballs. Even if you only fire one (low brightness), if it hits an electron, that electron is going to move.

Einstein had done the unthinkable. He had resurrected Newton's corpuscles. He declared that light was quantize. It was granular.

The Impossible Duality

This was a disaster for the scientific community.

"Wait a minute, Albert," they said. "We have proof light is a wave. We have the interference patterns. We have diffraction. A particle stream can't ripple!"

And Einstein essentially replied, "And yet, the Photoelectric Effect proves it comes in packets."

So, which is it? Is light a wave, or is it a particle?

The uncomfortable answer that slowly dawned on physicists was: **Yes.**

It is both. Depending on how you treat it, light can be a smooth, continuous wave, or it can be a jagged, hard little rock. It has a dual nature. It is schizophrenic.

But if you think that's confusing, it gets worse. Because physicists asked: "If light waves can act like particles, can solid particles act like waves?"

The Double Slit Experiment

This brings us to the single most famous, most repeated, and most mind-bending experiment in the history of science. If you ever feel like you understand quantum mechanics, revisit the **Double Slit Experiment**, and you will realize you don't.

Let's imagine we are back in Thomas Young's lab, but instead of light, we are using an electron gun. We are firing electrons—solid little chunks of matter—at a wall with two slits.

If electrons are particles (like marbles), they should fly through the left slit or the right slit and hit the back wall in two distinct bands.

We fire the gun. We look at the back wall. We don't see two bands. We see the **Interference Pattern**. We see the stripes of a wave.

Somehow, the electrons are behaving like water. They are going through the slits, rippling, interfering with each other, and landing in a pattern that only a wave can create.

"Okay," the physicists thought. "Maybe the electrons are bumping into each other like water molecules. Let's separate them."

They turned the gun down. Way down. They fired one single electron at a time. *Pop.* It hits the screen. One dot. *Pop.* Another dot.

They let this run for hours, firing one electron every minute. There is absolutely no way for these electrons to interact. There is only one in the universe at a time. It must choose: Left Slit or Right Slit.

But when they developed the picture on the back wall, the dots hadn't formed two bands. **They had formed the wave pattern.**

This is the moment logic breaks.

This implies that the single electron left the gun, traveled through space, approached the barrier, and then… split? Did it go through both slits at once, interfere with itself like a ghost, and then recombine to hit the wall?

How can a single object be in two places at once?

The Watcher

To figure this out, scientists decided to get clever. They put a sensor—a tiny "camera"—next to the slits to spy on the electron. They wanted to catch it in the act. They wanted to see which slit it really went through.

They turned on the electron gun. They turned on the sensor.

The first electron flew towards the slits. The sensor beeped: "It went through the Left Slit."

The next one flew. The sensor beeped: "Right Slit."

"Aha!" the scientists cheered. "It doesn't go through both! It chooses one!"

But then they looked at the back wall.

The wave pattern was gone. The interference stripes had vanished. Instead, there were just two boring, vertical bands of dots.

The moment they looked, the electron stopped acting like a wave and started acting like a solid particle.

It was as if the electron knew it was being watched. When unobserved, it behaved like a wave of infinite possibility, flowing through both doors at once. But the instant an observer tried to pin it down, it "collapsed" into a single, defined reality.

This is the **Measurement Problem**. It suggests that the observer is not external to the experiment. We are part of the system. We cannot observe the universe without changing it.

Einstein hated this. He hated the idea that reality was fuzzy until you looked at it. He wanted a universe that was solid and real whether he was watching it or not.

"Do you really believe," Einstein once asked a friend, "that the moon is not there when you are not looking at it?"

Quantum mechanics answers with a shrug: *"We can't be sure, Albert. We can't be sure."*

The Fog of War

By the mid-1920s, the world of physics had become a chaotic, fratricidal battlefield. The elegant, clockwork universe of Isaac Newton was gone, replaced by a confusing landscape where light could be a particle, particles could be waves, and nothing seemed to make sense anymore.

The old guard men like Einstein and Planck were desperately trying to find a way to make these new discoveries fit into the old rules. They wanted to visualize the atom. They wanted a picture in their heads. They imagined electrons orbiting the nucleus like tiny planets around a tiny sun. It was a comforting image. It felt like home.

But a new generation was rising. These were young, aggressive physicists, mostly in their twenties, who didn't care about "comforting images." They cared about the raw data. And leading this charge was a twenty-three-year-old German named Werner Heisenberg.

Heisenberg was a genius, but in the summer of 1925, he was a miserable genius. He was suffering from a catastrophic case of hay fever. His face was swollen, his eyes were shut tight, and he could barely think. Desperate for relief, he took a leave of absence from the University of Göttingen and fled to Helgoland a bleak, treeless hunk of red rock jutting out of the North Sea. There was no pollen there because there was almost no vegetation. Just rock, wind, and the ocean.

Isolating himself in that stark, alien landscape, Heisenberg decided to strip physics down to its bones. He decided that the problem with physics was that everyone was trying to *draw* the atom. They were trying to visualize electron orbits that no one had ever seen.

Heisenberg made a radical decision: *If we can't see it, we have no right to assume it exists.*

He abandoned the idea of orbits entirely. Instead, he focused purely on the math of the transitions the "jumps" electrons made when they released energy. Working feverishly through the night, often waiting for the sun to rise over the ocean, he developed a strange new type of calculus involving tables of numbers (matrices). It was abstract, ugly, and impossible to visualize. But when he finished the calculations at 3:00 AM one morning, the numbers matched the experimental energy levels of hydrogen perfectly.

Heisenberg had described the ghost, but in doing so, he had killed the ability to picture it.

The Death of Certainty

When Heisenberg returned to civilization, he began to realize the terrifying implication of his new math. It suggested that there was a fundamental limit to what human beings could know.

In the classical world the world of cars and baseballs measurement is passive. If you want to know where a car is and how fast it is going, you can use a radar gun and a camera. You can know both variables with perfect precision at the same exact moment. The car doesn't care that you are measuring it.

Heisenberg realized that in the quantum world, this is impossible.

He formulated an inequality that would become the bedrock of quantum mechanics: **The**

Uncertainty Principle.

It states that there are certain pairs of properties conjugate variables that are linked in a see-saw relationship. The most famous pair is **Position** (x) and **Momentum** (p).

Delta x Delta $p \geq h/2$

The equation tells us that the more precisely you know where a particle is (Delta x gets smaller), the less you know about where it is going (Delta p gets larger).

This is often explained to students using a clumsy analogy called the "Observer Effect." The story goes like this: *To see an electron, you have to bounce a photon off it. But because the electron is so tiny, the photon hits it like a bowling ball hitting a ping-pong ball. The act of measuring the position knocks the electron away, changing its speed.*

This analogy is useful, but it is technically a lie. It implies that the electron *has* a definite position and speed, and we are just too clumsy to measure it without touching it.

Heisenberg's discovery was far more profound and disturbing. He proved that the electron **does not possess** a precise position and momentum simultaneously. It is not a failure of our tools; it is a fundamental property of nature.

The Wave That Refuses to Be Pinned Down

To understand why, you have to remember that matter is a wave.

Imagine a sound wave. If I ask you, "Where exactly is the sound of that middle C?" you would look at me with confusion. A sound wave fills the room. It doesn't exist at a single "point." It has a wavelength (which gives it a pitch), but to have a wavelength, it must be spread out over space.

If you try to compress that sound wave into a tiny, single point to give it a perfect "position," you destroy the wave structure. You no longer have a frequency. You just have a sharp *pop*.

The universe forces a trade-off. You can know the location (the *pop*), or you can know the momentum (the *wave*). You cannot have both.

Killing Laplace's Demon

The philosophical fallout of this was nuclear.

For two hundred years, science had been built on **Determinism**. The French scholar Pierre-Simon Laplace had famously imagined an intellect "Laplace's Demon" that knew the precise position and momentum of every atom in the universe at a single instant. Laplace argued that for such a demon, "nothing would be uncertain and the future, as the past, would be present to its eyes."

If you know where everything is and how it's moving, you can predict the future perfectly. If the universe is a pool table, and you know the force of the break, you know where every ball will land. This meant that free will was an illusion; the firing of neurons in your brain was determined by the physics of the Big Bang.

Heisenberg took a knife to Laplace's Demon.

If you cannot know the position and momentum simultaneously, then you cannot predict the future. You can only predict **probabilities**. You can say, "The electron will *probably* be here," but you can never be sure.

Cause and effect were no longer ironclad chains. They were just betting odds.

Albert Einstein, watching this from Berlin, was horrified. He refused to accept a universe based on gambling. He launched a series of intellectual attacks on Heisenberg and his mentor, Niels Bohr, famously declaring, *"God does not play dice with the universe."*

Niels Bohr, tired of Einstein's stubbornness, eventually retorted: *"Einstein, stop telling God what to do."*

The Cat in the Iron Box

While Heisenberg was destroying determinism with his matrix math, another physicist was trying to salvage the situation. Erwin Schrodinger, an Austrian physicist with a flamboyant personal life and a taste for classical philosophy, hated Heisenberg's abstract, UN-visualizer math.

Schrodinger wanted to bring the "picture" back to physics. He sat down and developed his own theory, treating the electron not as a jumping particle, but as a physical wave that rippled around the nucleus. He produced the **Schrodinger Wave Equation**.

Physicists loved it. It felt like old-school physics. It was smooth. It was continuous. It didn't have those jerky, instant "quantum jumps." Schrodinger thought he had saved the day by proving that particles were actually just smeared-out waves of electric charge.

But then, a man named Max Born looked at Schrodinger's equation and ruined the party.

Born argued that the wave wasn't a physical wave of electricity. It was a **Probability Wave**. The height of the wave at any point didn't tell you how much "charge" was there; it told you the *odds* of finding the particle there if you looked.

This meant that before you look, the particle isn't in any specific place. It is in a state of **Superposition**. It is everywhere where the wave is greater than zero.

Schrodinger was disgusted. He had tried to fix the absurdity of quantum mechanics, and instead, he had given the world a tool that made it even weirder. He famously wrote, *"I don't like it, and I'm sorry I ever had anything to do with it."*

To show just how ridiculous this concept of "superposition" was, Schrodinger crafted a thought experiment. He wanted to take the microscopic weirdness of the atom and blow it up to the size of an animal, to show us how silly it sounded.

The Setup

Schrodinger asked us to imagine a steel chamber a bunker.

Inside, we place a cat. (Schrodinger had nothing against cats; he just needed a macroscopic biological subject).

Along with the cat, we place a "diabolical device." The device contains a tiny speck of radioactive substance, so small that in the course of one hour, there is a 50% chance one of the atoms will decay and release a particle, and a 50% chance it will not.

If the atom decays, a Geiger counter detects it. The counter triggers a relay that releases a hammer, which shatters a flask of hydroponic acid.

The logic is simple:

- Atom decays —> Poison released —> Cat dies.
- Atom does not decay —> Poison contained —> Cat lives.

We seal the box. We wait one hour.

Now, we ask the question: **Is the cat dead or alive?**

Common sense tells us that the cat is either dead or alive, regardless of whether we can see it. It is a binary state. The cat knows if it's alive. The universe knows.

But according to the strict interpretation of quantum mechanics (the Copenhagen Interpretation), we are wrong.

Because the radioactive atom is a quantum object, it is in a state of superposition until measured. It has decayed **AND** not decayed simultaneously.

But the cat is entangled with the atom. The cat's state is dependent on the atom's state. Therefore, according to the math, the cat enters a state of superposition.

Until we open the box and make an observation, the cat is mathematically, physically, and literally **both dead and alive at the same time.**

Collapsing Reality

Schrodinger's point was: "This is stupid." A cat cannot be dead and alive. Therefore, there must be something missing in our theory.

But history is full of irony. Instead of disproving the theory, Schrodinger's Cat became the mascot of quantum mechanics. It illustrated the central mystery of the field: **The Wave Function Collapse**.

The universe exists as a shimmering cloud of possibilities (the wave function). It contains all potential futures, all potential outcomes, overlapping each other. But when an observer interacts with the system when we open the box nature is forced to choose.

The wave function "collapses." The possibilities vanish. The ambiguity evaporates. The cat becomes dead OR alive.

This implies that we, the observers, are the architects of reality. We are the ones who force the universe to make up its mind.

The Many Worlds

There is, however, an alternative that is even more terrifying.

Decades later, a graduate student named Hugh Everett looked at Schrodinger's math and asked: "Why does the wave function have to collapse? What if it doesn't?"

Everett suggested that when you open the box, you don't force the universe to choose. Instead, the universe **splits**.

In one branch of reality, you open the box and find a live cat.

In another, equally real branch of reality, you open the box and find a dead cat.

Both outcomes happen. Both versions of *you* exist. This is the **Many-Worlds Interpretation**. It suggests that every time a quantum event has multiple outcomes, the universe cleaves in two. There are infinite versions of you, living out every possible life, spun off by the infinite decisions of subatomic particles.

Schrodinger just wanted to point out a flaw in the math. Instead, he opened a door to a multiverse where death itself might be just a matter of perspective.

Spooky Action at a Distance

By 1935, Albert Einstein had had enough.

He had spent thirty years watching the physics community take his beautiful, logical universe and turn it into a casino. He had accepted the quantization of light. He had begrudgingly tolerated Heisenberg's Uncertainty Principle. But there was a line he would not cross. He refused to believe in magic.

Einstein was a "Local Realist." This is a fancy way of saying he believed two things:

1. **Realism:** Objects have definite properties whether you are looking at them or not. The moon is there even if you close your eyes.
2. **Locality:** Objects can only influence each other if they touch, or if they send a signal to each other. Since nothing travels faster than light, there can be no instant communication across the universe.

If you wave your hand, the air moves. That moving air hits a leaf, and the leaf moves. That is "local" action. Cause and effect.

But quantum mechanics, in its growing arrogance, was predicting something else entirely. The equations suggested that two particles could become linked—**entangled**—in such a way that they stopped being two separate objects and became one single mathematical entity.

The theory claimed that if you took two entangled electrons and separated them by a galaxy, measuring one would *instantly* affect the other. No time delay. No signal. Just instant, telepathic obedience.

Einstein called this *"spukhafte Fernwirkung"*—**Spooky action at a distance.**

He decided it was time to kill quantum mechanics once and for all. He joined forces with two younger physicists, Boris Podolsky and Nathan Rosen, to write a paper that would expose the absurdity of this idea. It became known as the **EPR Paradox**.

The Case of the Quantum Gloves

To understand Einstein's argument, imagine a pair of gloves.

You have a Left Glove and a Right Glove. You put them in two identical boxes. You shuffle the boxes. You keep one box in New York and mail the other box to an astronaut on Mars.

Neither you nor the astronaut knows which glove is in which box.

You open your box in New York. You see a **Left Glove**.

At that exact moment instantly you know that the astronaut on Mars has the **Right Glove**.

Did you send a signal to Mars? No. Did you use magic? No. The astronaut has the Right Glove because *he always had the Right Glove*. From the moment you packed the boxes, the state was determined. The only thing that changed was your knowledge. This is "Hidden Variables." The reality was already there; you just didn't know it yet.

Einstein argued that entangled particles were just like gloves. If two electrons fly apart, one spinning "Up" and one spinning "Down," and you measure the first one as "Up," you instantly know the other is "Down."

Einstein said: *Obviously, the second electron was 'Down' the whole time. It didn't decide at the last second just because you looked at its brother. Physics is not telepathic. The particles carried 'hidden instructions' from the start.*

It was a brilliant argument. It saved common sense. It saved the speed of light. It made the universe logical again.

There was just one problem. Einstein was wrong.

The Bell Tolls

Einstein died in 1955, convinced he had won the argument. Bohr died a few years later. The debate seemed to be a philosophical stalemate.

But in 1964, a moody Northern Irish physicist named **John Stewart Bell** sat down and did something incredible. He figured out a way to test who was right.

Bell realized that the "Gloves" analogy and the "Quantum Magic" reality would actually produce slightly different statistical results if you measured them in a specific, tricky way (measuring the spin at different angles).

He created a mathematical litmus test called **Bell's Theorem**. It was a way to ask the universe: *"Are you per-determined (Gloves)? Or are you making this up as you go along (Magic)?"*

It took until the 1970s and 80s for technology to catch up, culminating in the experiments of Alain Aspect in Paris. They built the lasers. They fired the entangled photons. They ran the test.

The results were unequivocal. The universe failed the "Gloves" test.

Einstein's "Hidden Variables" did not exist. The particle on Mars did *not* have a defined spin before you measured the one in New York. It was in a haze of probability. The moment the exact nanosecond you measured the New York photon, the Mars photon snapped into the opposite state.

The universe is **Non-Local**.

The Cosmic Web

The implications of this are staggering. It means that space is not the barrier we think it is.

In our everyday lives, space separates things. If you are in the kitchen and I am in the living room, we are separate. I have to shout for you to hear me.

But in the quantum realm, space is just a backdrop. Two particles that have interacted are connected by a wave function that transcends distance. They are, in a very deep sense, the same object residing in two different places.

Some physicists speculate that because all matter in the universe was once crunched together into a singularity at the Big Bang, everything might be entangled with everything else to some degree. The atoms in your body interacted with the atoms in a star a billion years ago. Are there still threads connecting us?

We don't know. But we do know that Einstein's "spooky action" is real. We are currently using it to build the next generation of technology: **Quantum Cryptography**. We can send information that cannot be intercepted, because if a spy tries to look at the message, they collapse the wave function and destroy the data instantly.

Einstein wanted a universe of separate, distinct marbles. Instead, he got a universe that is a single, shimmering, interconnected web.

Walking Through Walls

If Entanglement is the mental ability of the quantum world, Tunneling is its physical superpower.

In the classical world the world of Isaac Newton barriers are absolute. Imagine you are standing at the bottom of a steep hill. You have a heavy bowling ball. You want to roll the ball over the hill to the valley on the other side.

You roll it. It goes halfway up, runs out of energy, stops, and rolls back down. You roll it harder. It goes 90% of the way up, stops, and rolls back down.

The rule is simple: If you do not give the ball enough energy to overcome the height of the hill ($E<V$), the ball will never, ever get to the other side. It doesn't matter if you try a million times. It doesn't matter if you wait a billion years. The hill is a wall.

But in the 1920s, physicists realized that subatomic particles don't care about hills.

The Ghost in the Wall

Remember that a particle is not a hard marble. It is a probability wave. When an electron approaches a barrier (like an electric field that repels it), the "wave" of the electron slams into the wall.

Most of the wave bounces back. This is reflection. The electron returns to where it came from. But and this is the key the wave function does not stop abruptly at the wall. It penetrates slightly. It has a "tail" that tapers off exponentially inside the barrier.

If the wall is thin enough, that tail might actually poke out the other side.

Since the wave function describes the *probability* of finding the particle, if a non-zero part of the wave exists on the other side of the wall, there is a non-zero chance that the particle will simply appear there.

It doesn't climb over the hill. It doesn't break the wall. It simply vanishes from one side and materializes on the other. It teleport.

This phenomenon is called **Quantum Tunneling**.

If you were a quantum object, you could run full speed at a brick wall, and every once in a while, you would pass right through it without a scratch. (Don't try this. You are made of trillions of atoms, and the odds of all of them tunneling at once are so low you would have to wait longer than the age of the universe).

But for single particles, it happens all the time. In fact, you owe your life to it.

The Sun Trap

For a long time, astronomers had a problem. They knew the Sun was powered by nuclear fusion smashing hydrogen atoms together to make helium.

To fuse two protons, you have to get them incredibly close together. But protons are positively charged. They hate each other. The electromagnetic force pushes them apart with tremendous power. This is the "Coulomb Barrier." It is a massive, steep energy hill.

To overcome this barrier and touch, the protons need to be moving fast. Very fast. This requires heat. When scientists calculated how hot the center of the Sun is, they got a figure of about 15 million degrees Celsius. That sounds hot, but according to classical physics, **it is not hot enough.**

At 15 million degrees, the protons are moving fast, but not fast enough to climb the Coulomb Barrier. They should run halfway up the hill and roll back down. No fusion. No energy. No sunlight.

The Sun should be a dead, dark ball of gas.

So why is it shining?

In 1929, physicist George Gamow applied the new quantum theory to the Sun. He realized that the protons didn't need to climb over the barrier. They could tunnel through it.

Because of their wave-like nature, every once in a while, a proton would defy the electromagnetic force, slip through the "wall" of repulsion, and fuse with its neighbor.

It is a rare event for an individual proton. But because the Sun has so many protons, it happens enough to release the blinding energy that lights our solar system. The sunlight hitting your face is a direct result of particles walking through walls.

The Touch of Death

But the Weaver's patterns are not always benevolent. The same mechanism that lights the stars also creates the most dangerous substance on Earth: **Radioactivity**.

Consider a Uranium atom. It is a monster a massive, unstable nucleus trembling with energy. It wants to spit out an "Alpha particle" (a chunk of two protons and two neutrons) to calm down.

But the Alpha particle is trapped inside the nucleus by the "Strong Nuclear Force." The walls of the prison are incredibly high. According to classical mechanics, the Alpha particle should never escape. Uranium should be stable forever.

But the Alpha particle keeps banging against the wall, trillions of times a second. Eventually, it gets lucky. It tunnels. It appears outside the nucleus and shoots away at high speed.

This is radioactive decay. It is the sound of prisoners escaping a quantum jail, one by one.

Taming the Ghost

Today, we don't just observe tunneling; we control it.

Reach into your pocket. Pull out your smartphone. Inside that phone is a flash memory chip. That chip stores your photos, your texts, and your operating system. It stores them in "Floating Gate Transistors."

To save a 1 or a 0, the chip needs to trap electrons in a microscopic cage, insulated by a layer of glass-like oxide. If the electrons are in the cage, it's a "0". If the cage is empty, it's a "1".

But the cage is sealed. How do we get the electrons in and out?

We zap them with a specific voltage that pushes them against the wall of the cage. We force them to tunnel. We make them walk through the insulator to get into storage, and we make them walk back out to erase the data.

Every time you save a selfie, you are utilizing a phenomenon that Isaac Newton would have considered witchcraft. You are forcing matter to teleport through solid barriers at your command. The ghost is no longer haunting us; it is working for us.

The Observer's Dilemma

We have now reached the point in our journey where physics stops sounding like science and starts sounding like a nervous breakdown.

We have established that particles exist in a state of superposition (a fuzzy cloud of possibilities) until they are measured. We have established that the act of measurement forces the wave function to "collapse" into a single, hard reality.

But this explanation, which is the standard "Copenhagen Interpretation" taught in universities for decades, has a massive, glaring hole in it. It is a hole so big that you can drive a truck full of metaphysics through it.

The question is: **What exactly is an observer?**

When we say "measurement causes collapse," who or what has the authority to measure?

Does it have to be a human with a PhD? Can a chimpanzee collapse a wave function? What about a cat? A bacteria? What about a Geiger counter left alone in a dark room?

This is the **Measurement Problem**, and it threatens to tear the objectivity of science apart. If reality requires an observer to define it, does the universe exist when no one is watching?

Wigner's Friend

In 1961, the Nobel Prize-winning physicist Eugene Wigner decided to push this problem to its breaking point. He created a variation of Schrodinger's Cat that introduced consciousness into the equation. It is known as the **Wigner's Friend Paradox**.

Imagine Wigner has a friend let's call him Bob. Wigner puts Bob inside a sealed laboratory. Inside that lab, Bob performs a quantum experiment. He measures a radioactive atom to see if it decays (Heads) or doesn't decay (Tails).

Wigner stands outside the locked door.

Inside the room, Bob looks at the atom. The wave function collapses. Bob sees a definite result: "Heads." He writes it down. For Bob, the superposition is over. Reality is solid.

But outside the door, Wigner has not yet looked. According to the strict rules of quantum mechanics, the entire laboratory including Bob is isolated from Wigner. Therefore, to Wigner, the system is still evolving.

Mathematically, Wigner must describe the room as a superposition. The state of the room is: **(Bob seeing Heads) + (Bob seeing Tails).**

To Wigner, Bob is currently suspended in a ghostly state of having seen both outcomes at once.

Then, Wigner opens the door. *Now* the wave function collapses for Wigner. He asks Bob, "What did you see?" Bob says, "I saw Heads! And I saw it five minutes ago!"

Here lies the paradox. Bob is convinced reality became definite five minutes ago. Wigner's math says reality only became definite *now* when he opened the door.

Who is right? If Wigner is right, then Bob was in a state of suspended animation until Wigner looked at him. This implies that there is a hierarchy of observers—that you need a "super-observer" to make the previous observer real.

But who observes Wigner? Does God have to look at the universe to make Wigner real?

The Consciousness Trap

This line of thinking led some brilliant physicists down a rabbit hole of mysticism. John von Neumann and Wigner himself flirted with the idea that **Consciousness** is the magic ingredient.

They suggested that atoms, rocks, and Geiger counters are all just physical matter—they obey the Schrodinger equation, so they can all become entangled and enter superposition. A Geiger counter can be "both clicked and clicked."

But the human mind? Consciousness? They argued that consciousness is something different. It is outside the physical system. It is the only thing capable of terminating the chain. They proposed that the wave function collapses only when the information hits a conscious mind.

This view puts humanity back at the center of the universe. It suggests that the cosmos was a Grey fog of probability for billions of years until the first eye opened and "collapsed" it into stars and galaxies.

It is a poetic idea. It is also scientifically hated, because it implies physics depends on psychology.

The Noise of the World

Fortunately, modern physics has found a way out of this solipsistic nightmare. It is called **Decoherence**.

The idea is that you don't need a conscious mind to collapse a wave function. You just need an environment.

A quantum system (like our atom) maintains its delicate superposition only as long as it is isolated. But the real world is noisy. It is filled with air molecules, heat radiation, and stray photons.

Imagine the atom as a spinning top, perfectly balanced. Now imagine throwing a handful of gravel at it. The gravel (the environment) hits the top, knocks it off balance, and forces it to fall over.

When a Geiger counter clicks, it moves a needle. That needle pushes air molecules. Those air molecules hit the wall. The wall warms up slightly. The information leaks out into the universe instantly. The "secret" is out. The superposition is destroyed not by a human eye, but by the irreversible interaction with the environment.

We don't see quantum weirdness in everyday life not because we are conscious, but because we are big. We are leaking data to the environment constantly. We are being "measured" by the air, the light, and the ground every nanosecond.

We are too loud to be quantum.

The Garden of Forking Paths

In 1957, a graduate student at Princeton named Hugh Everett III was drinking sherry with friends and thinking about Wigner's Friend.

He thought the whole debate about "collapsing the wave function" was messy. It required this arbitrary moment the "measurement" where the laws of physics suddenly changed from smooth waves to hard particles. Nature shouldn't be that inconsistent.

Everett had a radical, heretical idea. He asked a question that would alienate him from the scientific community for the rest of his life:

"What if the wave function never collapses?"

What if Schrodinger's equation applies to everything, all the time, forever?

The Universal Wave

Let's go back to Schrodinger's Cat. The cat is in a superposition: Dead + Alive.

In the standard view, when you open the box, you force nature to choose. Nature picks "Alive," and the "Dead" possibility vanishes from existence. It is deleted.

Everett said: *No. Physics doesn't delete things.*

He proposed that when you open the box, you become **entangled** with the cat. You don't force a choice. Instead, the universe splits.

The "timeline" branches into two distinct paths.

- **Path A:** There is a You who sees a live cat.
- **Path B:** There is a You who sees a dead cat.

Both outcomes happen. Both versions of "You" are equally real. Both versions feel like they are the "only" one. But they are now separated by the barrier of the multiverse. They can never interact again.

This is the **Many-Worlds Interpretation** (MWI).

The Vertigo of Infinity

This theory changes everything. It means that "probability" is an illusion.

If a weather forecaster says there is a 50% chance of rain, we usually think that means "it will either rain or it won't." Everett says: "It will rain, AND it won't."

Somewhere in the multiverse, you brought an umbrella. Somewhere else, you got soaked.

Every time a quantum event happens every time a radioactive atom decays, every time a photon hits a leaf the universe splits. This is happening trillions of times a second, in every atom of your body.

The universe is not a single timeline. It is a rapidly expanding fractal tree a garden of forking paths.

This implies that there are infinite versions of you living out every possible life consistent with the laws of physics. There is a universe where you didn't buy this book. There is a universe where you became a concert pianist. There is a universe where the asteroid missed the Earth 65 million years ago, and dinosaurs are reading this.

The Tragedy of Hugh Everett

You might think such a bold idea would make Everett famous. It didn't. It ruined him.

When he presented his thesis, the titans of physics specifically Niels Bohr hated it. They dismissed it as nonsense. They couldn't stomach the idea of infinite invisible universes just to solve a math problem.

Everett was crushed. He left theoretical physics entirely. He went to work for the Pentagon, doing mathematical modeling for nuclear war strategies. He became a chain-smoking alcoholic and died of a heart attack at age 51.

He died thinking his theory was a failure.

He didn't live to see the 21st century, where the Many-Worlds Interpretation has become one of the dominant theories in cosmology and quantum computing. David Deutsch, the father of quantum computing, famously said that quantum computers are proof of the multiverse because when a quantum computer solves a problem, it is effectively "outsourcing" the calculation to its shadow selves in parallel universes.

Quantum Immortality

There is one final, chilling thought experiment associated with Many-Worlds, one that Everett allegedly discussed in private.

Imagine the Russian Roulette version of Schrodinger's box. You sit in a chair with a gun pointed at your head, triggered by a quantum decay.

You pull the trigger.

- **Universe A:** The gun fires. You die.
- **Universe B:** The gun clicks. You live.

In Universe A, you are gone. Your consciousness ceases. But in Universe B, you are still there to observe the result. You hear the "click."

You pull the trigger again.

- **Universe B1:** Bang. Dead.
- **Universe B2:** Click. Alive.

From your perspective—from the inside of your own consciousness—you can *never* experience the timeline where you die. By definition, you are only aware of the branches where you survive.

Therefore, no matter how many times you pull the trigger, you will just hear *click, click, click*. You will watch friends and family die in other timelines, but *you* will seem to be miraculously lucky. You will survive accidents, diseases, and old age, trapped in an increasingly narrow sliver of probability where you simply continue.

This is **Quantum Immortality**. It is the ultimate terrifying consequence of the Weaver's infinite loom. It suggests that from your own subjective point of view, you might never be able to leave the stage.

The Standard Pattern

For thousands of years, humanity has been obsessed with a single question: "What is the bottom?"

If you take a piece of cheese and cut it in half, then cut that half in half, and keep going, do you eventually hit a layer of reality that cannot be cut? Or is it cheese all the way down?

The Greeks called the bottom "atomos" (uncut table). By the 1900s, we thought we had found it. We found the Atom. Then we split it and found the Nucleus and the Electron. Then we split the Nucleus and found Protons and Neutrons.

By the 1960s, physicists were feeling confident. They had a nice, tidy list of ingredients for the universe: Protons, Neutrons, Electrons, and Photons. Simple. Elegant.

But then they started building bigger hammers.

They built particle accelerators—massive underground rings designed to smash protons together at near light-speed. They wanted to see if the proton would break.

It did. And what came out was a mess.

The Particle Zoo

When they cracked the proton open, they didn't just find smaller Lego bricks. They found a chaotic explosion of weird, exotic particles that existed for only a fraction of a second. They found things they called "pins," "koans," "sigmas," and "deltas."

Physicists were horrified. They had wanted simplicity. Instead, they got a zoo. One physicist, Willis Lamb, famously joked, *"I used to think that when I discovered a new particle, I would be rewarded with a Nobel Prize. Now I think I should be punished with a fine."*

To make sense of this chaos, theorists like Murray Ell-Mann realized that all these new particles were actually just different combinations of even smaller, more fundamental ingredients. He called them **Quarks**.

(He took the name from a line in James Joyce's *Finnegans Wake*: "Three quarks for Muster Mark!" because they always seemed to come in groups of three).

This effort to organize the chaos resulted in the **Standard Model of Particle Physics**. It is the periodic table of the quantum world. And unlike the chemical periodic table with its 118 elements, the Standard Model is shockingly brief.

Everything you have ever touched, seen, or loved is made of just three things:

1. **Up Quarks**
2. **Down Quarks**
3. **Electrons**

That's it. A proton is two Ups and a Down. A neutron is two Downs and an Up. Stick them together, add an electron, and you have an atom.

There are other particles in the model (like the "Charm," "Strange," "Top," and "Bottom" quarks), but they are heavy, unstable ghosts that only exist in accelerators or exploding stars. The everyday universe is built on a budget.

The Forces (The Glue)

But a bag of bricks doesn't make a house. You need cement. You need forces to hold the quarks together.

In the quantum world, forces aren't magic fields. They are things you throw back and forth. Imagine two skaters on an ice rink throwing a heavy medicine ball between them. The act of throwing and catching pushes them apart. That is a repulsive force.

These "medicine balls" are called **Bostons**. They are the messenger particles of the universe.

- **The Photon:** The messenger of Electromagnetism. It tells electrons to repel each other.
- **The Gluon:** The messenger of the Strong Nuclear Force. It "glues" quarks together inside the proton. It is so strong that if you try to pull two quarks apart, the energy you use creates *new* quarks out of thin air before the bond breaks.
- **The W and Z Bostons:** The heavy, sluggish messengers of the Weak Force (responsible for radiation).

The God Particle

There was just one problem with this beautiful model. The math said that all these particles should be weightless. According to the equations, electrons and quarks should be zipping around at the speed of light, never slowing down to form atoms. The universe should be a soup of radiation.

But obviously, we have mass. We are heavy. Something is slowing us down.

In the 1960s, a shy Scotsman named Peter Higgs proposed a solution. He suggested that the entire universe is filled with a thick, invisible energy field—like a cosmic molasses.

When a particle tries to move through this field, it experiences drag. It wades through the molasses. This resistance is what we call "Mass."

- A photon cuts through the field like a needle; it doesn't interact, so it has no mass and travels at light speed.
- An electron interacts a little bit; it gets a little heavy.
- A Top Quark interacts heavily; it gets bogged down and becomes massive.

This field became known as the **Higgs Field**. And if it existed, it had to have its own particle: **The Higgs Boston**.

For 50 years, the Higgs Boston was the holy grail of physics. We built the Large Hadron Collider (LHC)—a 17-mile ring beneath the border of France and Switzerland—specifically to find it.

On July 4, 2012, the announcement was made. They had found a bump in the data at 125 GeV. They had found the Higgs. The audience in the auditorium wept.

We had found the final tile in the mosaic. We now knew what the Weaver used to give substance to the universe.

The New Machine

While particle physicists were smashing the universe apart to see what it was made of, another group of scientists was asking a different question: "Can we build a computer out of this?"

For the last 50 years, the digital revolution has been powered by the transistor. A transistor is just a switch. It is either **ON (1)** or **OFF (0)**. This is the **Bit**. Everything your laptop does—from playing a movie to calculating a spreadsheet—is just a massive collection of 1s and 0s flipping back and forth very fast.

But as we shrink transistors down to the size of atoms, we run into a problem: Quantum Tunneling. Electrons start jumping across the switches. The classical world ends.

Instead of fighting this, physicists decided to harness it. They proposed a new type of computer based not on bits, but on **Qubits** (Quantum Bits).

The Power of "Both"

A classical bit is like a coin glued to the table. It is either Heads (1) or Tails (0). A Qubit is like a spinning coin. While it is spinning, it is a blur of Heads and Tails at the same time. It is in **Superposition**.

This sounds like a minor difference, but the math is explosive.

- If you have 2 classical bits, you can store one of four combinations (00, 01, 10, 11).
- If you have 2 Qubits in superposition, you store **all four** combinations simultaneously.

If you add more Qubits, the power grows exponentially.

- 3 Qubits = 8 states at once.
- 300 Qubits = More states than there are atoms in the observable universe.

Imagine you are looking for a specific book in a library with a million books.

- A **Classical Computer** is like a super-fast librarian who runs down the aisle checking one book at a time. *No. No. No. Yes.* It takes time.
- A **Quantum Computer** creates a million parallel versions of the librarian. They check every single book in the library at the exact same moment.

This is **Quantum Parallelism**. It allows us to solve problems that would take a normal supercomputer millions of years.

The Entangled Brain

But superposition is only half the magic. The other half is **Entanglement**.

In a quantum computer, the Qubits are entangled with each other. If you flip one Qubit, it affects the others. This allows the computer to process complex patterns and correlations that classical computers simply miss.

We process the data by running "Quantum Gates." These aren't like logic gates in a normal chip. They are microwave pulses or laser beams that rotate the probability clouds of the Qubits. We choreograph a dance of probabilities so that the wrong answers destructively interfere (cancel out) and the right answer constructively interferes (amplifies).

When we finally measure the Qubits at the end of the calculation, the wave function collapses, and the correct answer pops out.

The Stakes

Why does this matter? Why are Google, IBM, and governments pouring billions into this?

1. **Medicine:** Classical computers are terrible at simulating chemistry because molecules are quantum objects. To simulate a caffeine molecule perfectly, you would need a computer the size of the Earth. A quantum computer can simulate molecules naturally. It could let us design new drugs, new batteries, and new materials in days instead of decades.
2. **Encryption (The Apocalypse):** Almost all global security (your bank account, military secrets) relies on the fact that it is hard to factor large numbers. A classical computer would take a trillion years to crack 256-bit encryption. A powerful quantum computer running "Short's Algorithm" could do it in hours. This is why there is a race to build "Quantum-Safe" internet protocols.
3. **Artificial Intelligence:** Quantum computers could search through massive datasets and optimize neural networks at speeds we can't comprehend, potentially unlocking true AI.

Quantum Supremacy

In 2019, Google claimed to have achieved **"Quantum Supremacy."** Their processor, "Sycamore," performed a calculation in 200 seconds that they claimed would take the world's fastest supercomputer 10,000 years to finish.

It was a useless calculation generating random numbers but it was the Wright Brothers moment. The plane had flown. It only flew 120 feet, but it proved flight was possible.

We are now in the era of the "NI SQ" (Noisy Intermediate-Scale Quantum). Our Qubits are fragile. They collapse if you look at them wrong. Cosmic rays disrupt them. We are trying to build a symphony in the middle of a hurricane.

But the machine is being built. The Weaver is being mechanized. We are learning to compute with the fabric of reality itself.

The Final Thread

We have come a long way. We have pix elated energy, split the atom, tangled the threads of space, and even built computers out of ghosts. We have a nearly perfect manual for how the universe works.

We have **Quantum Mechanics** for the very small (atoms, subatomic particles). It works beautifully. It predicts experiments to the tenth decimal place. We have **General Relativity** for the very big (stars, galaxies, gravity). It also works beautifully. It predicts the orbits of planets and the bending of light by black holes.

There is just one problem. These two theories hate each other.

They are like two families living in the same house who refuse to speak.

- **General Relativity** (Einstein's masterpiece) says that space is a smooth, continuous fabric. It curves gently like a rubber sheet under a bowling ball. It is deterministic.
- **Quantum Mechanics** says that the world is discrete, pix elated, and frantic. It says space is a chaotic, jittery foam of particles popping in and out of existence. It is probabilistic.

Most of the time, they ignore each other because they live in different neighborhoods. Atoms don't have enough gravity to care about relativity.

Stars are too big to care about quantum uncertainty.

But there are places in the universe where they are forced to meet. Places where there is a massive amount of stuff crushed into a tiny space.

1. ** The Center of a Black Hole.**
2. ** The Moment of the Big Bang.**

In these places, you need both theories. You need to calculate the gravity of a star (Relativity) acting on a point smaller than an atom (Quantum).

When physicists try to combine the equations, the math doesn't just fail; it explodes. The equations start spitting out "Infinity" as an answer for everything. Infinite density. Infinite curvature. Infinite energy. In physics, "Infinity" means "You are wrong."

This is the greatest crisis in modern science. We are missing the bridge. We are missing the **Theory of Everything** (Quantum Gravity).

The Symphony of Strings

For decades, the leading candidate to fix this broken marriage has been **String Theory**.

The idea is beautiful in its simplicity. We used to think electrons and quarks were zero-dimensional points little dots with no size. String theory suggests that if you zoomed in on an electron, you wouldn't see a dot. You would see a tiny, vibrating loop of energy a string.

Think of a guitar string. If you pluck it one way, it makes an 'A' note. Pluck it another way, it makes a 'C'. In String Theory, the "note" is the particle.

- Vibrate the string one way, it looks like an **Electron**.

- Vibrate it another way, it looks like a **Quark**.
- Vibrate it a third way, and it looks like a **Gravitation** (the particle of gravity).

This solves the math problem! Because strings have a physical length, they "smear out" the interactions. They prevent the two theories from clashing at a single zero-point. The infinities disappear. Gravity finally fits into the quantum puzzle.

But there is a catch. A big one. For the math of String Theory to work, the strings need room to vibrate. They need more directions than Up-Down, Left-Right, and Forward-Backward. The math requires **10 or 11 dimensions**.

Where are these extra 6 or 7 dimensions? Theorists say they are "curled up" (compacted) so tightly that we can't see them. Imagine looking at a garden hose from an airplane. It looks like a 1D line. But if you are an ant walking on the hose, you realize it has a second dimension circumference. You can walk *around* it. String theorists believe every point in our 3D space is actually a tiny, knotted ball of 6 extra dimensions that we are too big to perceive.

The Weave of Space

Not everyone buys String Theory. It has never been proven experimentally. It might just be beautiful math with no reality.

The main rival is **Loop Quantum Gravity (LQG)**. Proponents of LQG say, "Forget the extra dimensions. The problem is space itself."

Relativity assumes space is smooth. LQG says: **No.** It suggests that space is not a background stage. Space is made of "chunks." Space itself is quantize. Imagine a chain-mail armor. From a distance, it looks like smooth steel fabric. Up close, it is made of interlocking rings. LQG says space is a network of tiny loops. There is a smallest possible unit of space (the Planck Volume). You cannot divide space smaller than this.

If this is true, the universe is like a giant Lego set. The "singularity" of a Black Hole can't be infinitely small because you can't crush the Lego down to zero. The collapse stops at the Planck size. The math is saved without needing invisible dimensions.

So, which is it? Are we made of vibrating strings in 11 dimensions? Or are we living in a chain-mail universe? We don't know yet. The Weaver has one final knot that we haven't untangled.

Epilogue

The Unfinished Pattern

We started this book by looking at your hand. We zoomed in past the skin, past the cells, past the molecules, and into the atom.

We found that the solidity of the world is a lie. We found that matter is mostly empty space, held together by forces that act like ghosts. We learned that:

- Energy comes in pixels (Quanta).
- Particles are waves of probability until you look at them.
- Objects can be in two places at once.
- Separated twins can communicate instantly across galaxies.
- Reality might be branching into infinite multiverses every second.

It is easy to feel small when faced with this. It is easy to feel like the universe is a cold, chaotic storm where common sense goes to die.

But there is another way to look at it.

For centuries, we treated the universe like a machine. We thought it was a clock, grinding away without soul or purpose. We thought we were just cogs. Quantum mechanics killed the machine.

The universe it reveals is not a clock. It is flexible. It is participatory. It requires observation. It suggests that the boundary between "out there" (the physical world) and "in here" (the observer) is not as thick as we thought.

Sir James Jeans, the great physicist, once wrote:

> "The universe begins to look more like a great thought than like a great machine."

We are not just walking through the Weaver's tapestry. We are the threads. We are the way the universe observes itself. The atoms in your eyes were forged in stars billions of years ago, and now, after drifting through the cosmos, they have come together to form a brain that is capable of thinking about Quantum Mechanics.

The mystery is not solved. The Theory of Everything remains out of reach. But that is the beauty of science. If we knew everything, the story would be over. The Weaver is still working. The pattern is still growing. And the best part?

We get to watch.

www.ingramcontent.com/pod-product-compliance
Lightning Source LLC
LaVergne TN
LVHW090432210926
848480LV00002B/791
9798233447594